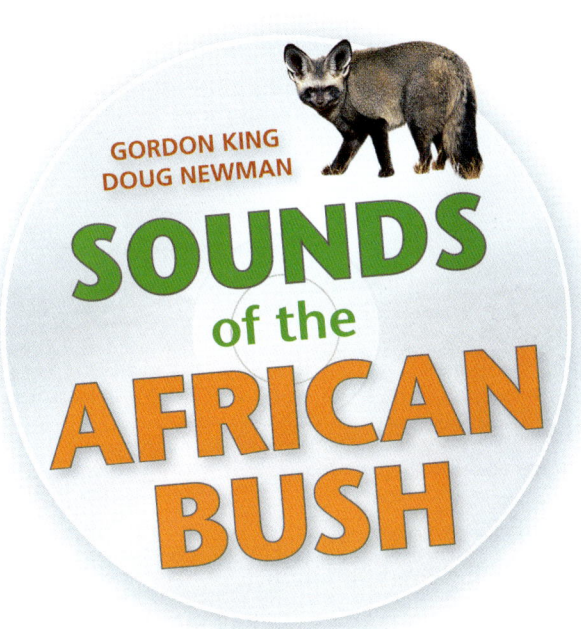

This book is dedicated to Sarah and Karen.

Grateful thanks to Chris and Mathilde Stuart
for permission to use their mammal distribution maps.

Published by Struik Nature
(an imprint of Penguin Random House (Pty) Ltd)
Reg. No. 1953/000441/07
The Estuaries No. 4, Oxbow Crescent, Century Avenue, Century City 7441
PO Box 1144, Cape Town, 8000 South Africa

Visit www.randomstruik.co.za and join the Struik Nature Club
for updates, news, events, and special offers.

First published in 2013
3 5 7 9 10 8 6 4

Copyright © in text, 2013: Doug Newman and Gordon King

Copyright © in maps, 2013: Chris & Mathilde Stuart (species 1–37), Louis Du Preez (species 72–76),
Penguin Random House (Pty) Ltd (species 38–69)

Copyright © in photographs, 2013: Book: As listed next to each photograph; IOA=Images of Africa.
Front cover: *Top row, from left:* Hein von Horsten, Mike Picker, Doug Newman, Nigel Dennis/IOA. *Deep etches:*
Gordon King (top), Louis du Preez (bottom left), Nigel Dennis/IOA (bottom right).
Back cover: *Top row, from left:* Nigel Dennis/IOA, Peter Steyn, Nigel Dennis/IOA, Nigel Dennis/IOA. *Deep etches:*
Doug Newman (right), Gordon King (left). *Bottom row, from left:* Louis du Preez, Nigel Dennis/IOA.
CD Cover: *Row, from left:* Nigel Dennis/IOA, Louis du Preez, Nigel Dennis/IOA. *Deep etches:* Daryl & Sharna
Balfour/IOA (left), Doug Newman (right), Nigel Dennis/IOA (bottom).

Copyright © in animal call recordings, 2013: Clem Haagner (tracks 1–3, 5–28, 30–37, 69–71),
Doug Newman (tracks 4, 29, 38–57, 59–68, 73), Gordon King (track 58), Louis du Preez (tracks 45, 72, 74–76)

Copyright © in published edition, 2013: Penguin Random House (Pty) Ltd

Publisher: Pippa Parker
Managing editor: Helen de Villiers
Editor: Lisa Delaney
Design director: Janice Evans
Typesetter: Neil Bester

Reproduction by Hirt & Carter Cape (Pty) Ltd
Printed and bound by Toppan Leefung Packaging and Printing (Dongguan) co., Ltd, China

All rights reserved. No part of this publication may be reproduced, stored in
a retrieval system, or transmitted, in any form or by any means, electronic,
mechanical, photocopying, recording or otherwise, without the prior
written permission of the copyright holder(s).

ISBN: 978 1 92057 241 9

CONTENTS

Introduction	4
How to use this book	5
Descriptions and images of the 76 animals featured on the accompanying CD	6
Index	48

Please listen responsibly!
Animals may be disturbed by the sound of their own calls or by the calls of other species. If possible, please do not play these calls in the wild, as they may interfere with animals' behaviour.

INTRODUCTION

Sounds from the African bush are complex and diverse: a chorus of chirps, grunts, croaks and growls. The calls vary widely, but many evoke wild Africa, from the African Fish Eagle's plaintive cry to the lion's resonant roar and the hyaena's eerie whoop.

Animals use sounds for a variety of reasons, from defending a territory to giving warnings, attracting a mate, locating food and communicating over distance. The breeding season brings an increase in vocalisations for almost all animal species. These sounds are usually strongly tied to mating rituals, displays and the proclamation of territory.

Knowing basic animal life histories (e.g. feeding time and mating season) will help in the identification of a species based on its calls. For example, many predatory animals, like owls and cats, are typically heard at night as they hunt for food, while most birds are active during the hours around sunrise and sunset.

In general, birds tend to have beautiful, melodious and varied songs (though some can make harsh or grating sounds). Frogs have a diverse array of calls, ranging from croaks to high-pitched whistles – some of which even sound similar to bird song.

Insects are a diverse group, with an equally diverse range of sounds, though they often produce shrill and high-pitched sounds. Mammal calls are also highly variable, but are dominated by loud grunts and roars that can often travel long distances.

With more than 950 bird species, 200 mammal species and myriad frog and insect species, a comprehensive collection of African wildlife calls would be vast. The aim of this book is thus to provide a general introduction to the calls of the most commonly heard species within the context of an African game reserve.

It is thought that elephants can communicate over very long distances using ultra-low frequencies.

HOW TO USE THIS BOOK

- This book, together with the CD, will help you to recognise some of the amazing sounds made by our African fauna. The book features a photograph and information on 76 species likely to be encountered in the African bush. A map within each entry shows where in Africa the animal occurs.

- Play the CD to listen to the calls of the species in this book. Use the CD in conjunction with the book to see photographs of the animals as you learn about them.
- Once you learn to recognise the calls of the bush, you will be able to build up a picture of the nearby animals, and perhaps even spot them.

CD track number and species' common and scientific names.

Distribution map showing where in sub-Saharan Africa the animal is found.

Habitat and dietary information indicated for each species.

Information on the call(s) and the corresponding track number.

20 | Mohol Galago (Lesser Bushbaby)
Nagapie
Galago moholi

Known for their ability to jump great distances, mohol galagos move about the tree canopy with ease.

Although these nocturnal animals live in small family groups, they tend to forage alone. Most foraging occurs in trees, but they will sometimes forage on the ground.

Mohol galagos are common throughout their range, and can be found from the southern parts of central Africa to the northeastern parts of southern Africa.

HABITAT Woodland (acacia, savanna and riverine).

DIET Mostly insects and tree gum; also other small invertebrates.

TRACK N° 20

A squeaky, high-pitched, baboon-like barking and very loud, siren-like wails. These nocturnal sounds can carry long distances.

21 | Samango (Sykes') Monkey
Samango-aap
Cercopithecus albogularis

Samango monkeys live in large troops of up to 40 members, which are typically led by one dominant male.

These monkeys spend most of their time in trees, though they will occasionally forage on the ground.

Although more reserved than many monkey species, samango monkeys have a range of loud calls, which makes them easy to locate in their dense habitats.

They have a patchy distribution that ranges from the east coast of South Africa up to Somalia and into parts of central and eastern Africa.

HABITAT Various, from swamps to scrub, though always densely forested.

DIET A variety of plant matter, including fruits and flowers; insects may also be eaten.

TRACK N° 21

Loud honking and barking calls, often with a guttural tone, may be heard during the day.

Adult female   *Adult male (with characteristic mane)*

1 | Lion
Leeu
Panthera leo

These are highly social animals that live in family groups known as prides. Although females make up the core of a pride, a single, dominant male acts as leader. When a new male takes over a pride, he will often kill the offspring of other males in order to bring the females rapidly into oestrus and mate with them, so ensuring that his genetic line predominates in the pride.

Lions tend to hunt at night or during the hours around sunrise and sunset; they typically rest during the day. Lions are not particularly timid animals and are often seen resting under trees in large family groups.

They are patchily distributed throughout sub Saharan Africa, but are absent from most of central western Africa and much of southwestern Africa. In many regions, lions are restricted to protected areas.

HABITAT Adaptable to a wide range of habitats, but tend to avoid mountainous regions and thick forests. Although often found in arid regions, they also avoid true desert.

DIET Mostly medium to large mammals, typically wildebeest, impala, zebra, buffalo and warthogs. They may opportunistically consume smaller prey, such as birds, rodents and even insects. Females do the bulk of the hunting as a co-operative group, though male lions may hunt as well.

TRACK N° 01

A loud, booming roar may be heard during the night. They also make other low-pitched purring and grunting sounds.

Leopards are good climbers and are sometimes seen resting in trees

2 | Leopard
Luiperd
Panthera pardus

Outside of the breeding season, leopards lead a solitary existence; groups one may see are likely to be mothers with their cubs.

They are excellent climbers and will often rest in trees, where they are less likely to be disturbed. In some cases, they may even drag their prey up into trees to feast without interruption.

Leopards are mostly nocturnal, but may also be active at sunrise or sunset.

They are relatively numerous in sub-Saharan Africa, and are the most widespread of the large African cats. Leopards are adaptable to a wide range of environments: they are not wary of human settlements and have even been spotted on the fringes of large cities.

HABITAT Varies widely. They are often associated with rocky and hilly areas, including forests and semidesert. In order to hunt, they require some form of cover, such as trees, rocks or thick bushes.

DIET A wide range of prey animals, predominantly small to medium-sized antelope. Also very small mammals, such as dassies and rabbits. Occasionally, they may also eat birds, fish, or even fruit. Leopards are among the few predators known to attack porcupines.

 TRACK N° 02

A soft, but harsh, breath-filled snarl (similar to the sound of sawing wood) may be heard at any time of the day or night.

Cheetahs have distinctive black 'tear' markings

3 | Cheetah
Jagluiperd
Acinonyx jubatus

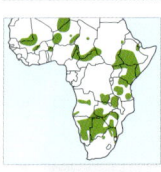

With a lighter build than that of the other large cats, cheetahs are the fastest mammals on Earth. From rest, they can reach speeds of more than 100 km/hour in a period of about 3 seconds, though these speeds can only be sustained for a few hundred metres.

When resting, they often lie atop a raised vantage point, which enables them to watch for prey or advancing danger.

Unlike other cats, cheetahs are unable to retract their claws completely, making their tracks easily identifiable. Their facial 'tear' marks are also a unique feature among the large cats.

Cheetahs have a patchy distribution throughout sub-Saharan Africa, and are absent from much of central western Africa. They are more prevalent in southern and eastern Africa, with the exception of South Africa, where they are relatively uncommon.

HABITAT   Open areas, such as grassland and desert plains, where it is easier for them to hunt.

DIET   Mostly small to medium-sized mammals, consisting largely of antelope; also birds of various sizes, including ostriches.

TRACK N° 03

From soft purring to a loud, almost bird-like chirp; may be heard at any time of the day or night.

4 | African wild dog
Wildehond
Lycaon pictus

Also known as the Cape hunting dog, the African wild dog is a threatened species. Reintroduction into smaller reserves has been largely unsuccessful due to conflict with lions.

These are highly social animals that hunt collectively as a pack (usually 10–15 individuals), chasing prey until it tires.

African wild dogs are predominantly found in the protected areas of southern and eastern Africa. They have a very fragmented distribution.

HABITAT Open habitats and occasionally woodland; they avoid forests.

DIET Medium-sized to large antelope. They may also hunt smaller prey, though tend to prefer large animals that can be shared with the pack.

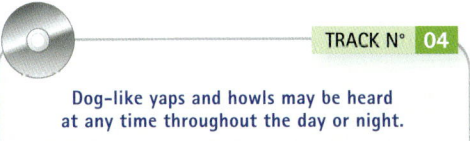

TRACK N° 04

Dog-like yaps and howls may be heard at any time throughout the day or night.

5 | Black-backed jackal
Rooijakkals
Canis mesomelas

These jackals will often venture close to settlements, frequenting farms and smallholdings.

In areas populated by humans, they often adopt nocturnal habits. However, in more remote areas, they are most active around sunrise and sunset – when temperatures are cool, but there remains enough light by which to hunt.

Black-backed jackals are widespread throughout eastern and southern Africa.

HABITAT Found in most habitats, with the exception of thick forests.

DIET Best known for scavenging carrion, though will also hunt small creatures, including insects, birds and rodents; they have also been known to eat fruit.

TRACK N° 05

A shuddering, almost wolf-like howl with an eerie quality, which may be heard at dawn, dusk or during the night.

6 | Side-striped jackal
Witkwasjakkals
Canis adustus

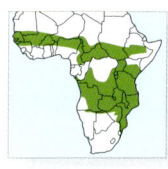

Side-striped jackals are predominantly nocturnal and often solitary, although territories are thought to be held by mating pairs.

As with black-backed jackals, they may conflict with humans in rural farming areas.

Side-striped jackals are found throughout much of central Africa, with the exception of dense forests in western Africa. They are absent from most of southern Africa.

HABITAT Typically woodland, though they are sometimes found in open habitats.

DIET A wide range of smaller prey, including small mammals, birds and insects, as well as carrion; they have also been known to eat fruit, grass and even crops.

TRACK N° 06

A high-pitched, eerie sound consisting of descending whistles and short, single-whistled notes; heard at dawn, dusk or night.

7 | Bat-eared fox
Bakoorvos
Otocyon megalotis

Although mainly nocturnal, bat-eared foxes are occasionally active during the day, particularly during the winter months.

They are often found in small family groups, consisting of a breeding pair (which mates for life) and its offspring. They may temporarily gather in larger numbers at rich food sources.

Bat-eared foxes have a playful nature and are often seen romping.

They occur in eastern Africa and across most of southern Africa.

HABITAT Bare, open ground or grassland with short grasses, though may extend into woodland with bare patches.

DIET Mainly termites: thus termite mounds are a crucial habitat component. They will also eat insects, rodents and birds.

TRACK N° 07

A jackal-like, snarling bark with an eerie quality, which may be heard at any time. The young are the most vocal.

8 | Spotted hyaena
Gevlekte hiëna
Crocuta crocuta

9 | Brown hyaena
Bruin hiëna
Parahyaena brunnea

Spotted hyaenas may be active at any time, but are less commonly seen during the day.
 They are typically found in family groups, led by a dominant female. These groups, consisting of 15 or more individuals, will aggressively defend their territory.
 Spotted hyaenas occur throughout most of sub-Saharan Africa, but are absent from parts of central western Africa and much of South Africa.

HABITAT Varies widely, though dense forest and desert are unsuitable.

DIET Medium-sized to large ungulates. Although capable hunters, they will often scavenge.

Although they forage alone, brown hyaenas will often share their territory with extended family members.
 All members of a family unit will take part in raising cubs.
 Brown hyaenas are usually nocturnal scavengers, and are known to bury excess food for future consumption.
 They are almost entirely restricted to the northern parts of southern Africa.

HABITAT A wide range, from coastal areas to desert and into savanna woodland.

DIET Almost exclusively scavengers, though they may hunt smaller mammals and invertebrates; they occasionally eat fruit.

TRACK N° 08

Nocturnal laughing, chattering and whooping sounds; best known is the rising whoop, which has a ghostly quality.

TRACK N° 09

Nocturnal guttural snarls and soft dog-like yelps; cubs make harsh whining noises when begging for food.

10 | Honey badger
Ratel
Mellivora capensis

The honey badger's common name is derived from its tendency to seek out beehives, which it raids for honey and bee larvae.

Honey badgers have a well-deserved reputation for being ferocious and, despite their relatively small size, can be very aggressive when cornered.

These are solitary animals that tend to exhibit mostly nocturnal habits, though they may hunt at any time of the day or night.

Honey badgers are found throughout most of sub-Saharan Africa, except along the coastal strip of the Namibian desert.

HABITAT Various, with the exception of deserts.

DIET Typically invertebrates and rodents, though may include fruit and carrion.

 TRACK N° 10
Growls and grunting noises, typically heard at night. They make a nasal, laughing sound when disturbed.

11 | Cape porcupine
Ystervark
Hystrix africaeaustralis

When threatened, porcupines will raise their quills towards the threat and rush sideways or backwards, in an attempt to stab their attacker. They inhabit disused aardvark burrows, which they modify to have many chambers and entrances, or they shelter in rock crevices and caves.

Porcupines are solitary, nocturnal foragers that take refuge in their burrows during the day.

Southern African porcupines are found throughout central and southern Africa.

HABITAT Highly variable, including forest, grassland, desert and mountainous areas.

DIET Mainly bulbs and roots, but also vegetables, tree bark and, occasionally, bones (for mineral content).

 TRACK N° 11
A variety of growls when threatened. When attacked, a rattling sound is made using their hollow tail quills. Only heard at night.

12 | Yellow mongoose
Witkwasmuishond
Cynictis penicillata

Yellow mongooses live in family groups that are centred around a single breeding pair.
These highly social animals may share their burrows with other species, typically meerkats and ground squirrels. However, despite their social behaviour, they are often seen foraging alone.

Yellow mongooses are restricted to southern Africa, but are relatively widespread across this region; within their range, these are the mongooses most likely to be encountered in the wild.

HABITAT Dry grassland and semi-arid scrub.

DIET Mostly insects and other invertebrates; also small vertebrates and carrion.

TRACK N° 12

A collection of soft growls and purrs may be heard during the day.

13 | Banded mongoose
Gebande muishond
Mungos mungo

Banded mongooses can easily be distinguished from other mongooses by the characteristic stripes across their back.
They are highly social animals that live in troops of up to 30 individuals.

During the day, banded mongooses will forage in loose groups, all the while maintaining vocal contact with fellow members.

They are found throughout much of central and eastern Africa, extending into the northern parts of southern Africa; they are absent from the densely forested parts of western Africa.

HABITAT Savanna and woodlands located near good sources of drinking water.

DIET Typically insects and small birds; reptiles and amphibians are sometimes consumed.

TRACK N° 13

During the day, a collection of soft growls, squeals, high-pitched yelps and assorted purrs may be heard.

14 | Slender mongoose
Swartkwasmuishond
Galerella sanguinea

Although not typically social animals, slender mongooses may share their dens with other species.
  They are diurnal, but may hunt on moonlit nights. Slender mongooses are capable climbers, and are able to hunt birds and other tree-dwellers, although they themselves are primarily ground-dwellers.
  They are found throughout sub-Saharan Africa except in the dense forests of central western Africa, extending down across southern Africa, but are absent from coastal Namibia, southerly South Africa and the dry Horn of Africa.

HABITAT Savanna and semi-arid plains; absent from desert regions and very dense forest.

DIET Various invertebrates and small vertebrates; they are able to kill venomous snakes, which they sometimes eat.

   TRACK N° **14**

A collection of soft growls and purrs may be heard during the day.

15 | Meerkat (Suricate)
Stokstertmeerkat
Suricata suricatta

Meerkats are known for their tendency to stand on their hind legs as they search for danger – as do other mongoose species, but not to the same degree as meerkats.
  These highly social animals live in groups of up to 40 members, which are typically led by a dominant breeding pair.
  When foraging, meerkats maintain vocal contact with group members.
  They live in the drier regions of southern and western southern Africa, with a narrow band extending up into Lesotho, but are absent from the extreme coastal areas.

HABITAT Arid habitats, including some types of grassland; they avoid treed areas.

DIET Insects and other invertebrates; also small vertebrates and eggs.

   TRACK N° **15**

During the day, a collection of soft growls and purrs may be heard.

16 | Red bush squirrel
Rooi eekhoring
Paraxerus palliatus

17 | Tree squirrel
Boomeekhoring
Paraxerus cepapi

Although red bush squirrels are usually solitary, small groups, typically consisting of a breeding pair and their young, may be seen in loose association.

They are highly vocal, diurnal animals that are often found foraging in thick understorey.

As a result of habitat destruction, red bush squirrels have become endangered.

Their range extends along the east coast of Africa, from Somalia to South Africa, and inland to the Great Rift Valley and parts of Tanzania.

HABITAT Forests, montane forests and scrublands.

DIET A wide range of plant matter, including fruit and roots; also lichens and insects.

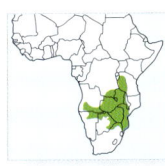

Tree squirrels are the most abundant squirrels in southern Africa, and are often found in close proximity to human settlements.

Although mostly solitary, small groups may maintain loose associations.

Tree squirrels are highly vocal, diurnal animals, which can be seen foraging on the ground or in trees. They will often hide their food for future consumption.

They are found in the northern parts of southern Africa, extending into the eastern parts of central Africa.

HABITAT Preferentially woodland, but will sometimes venture into more open savanna.

DIET Fruit, nuts and occasionally insects.

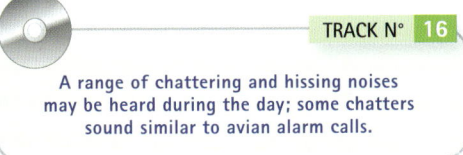

TRACK N° 16

A range of chattering and hissing noises may be heard during the day; some chatters sound similar to avian alarm calls.

TRACK N° 17

Vocal only during the day, they produce an agitated, chattering laugh with clicking and hissing.

18 | Dassie (Rock hyrax)
Klipdas (Dassie)
Procavia capensis

Dassies are social animals that live in family groups, led by a dominant male and female pair.
They are diurnal and are most often seen basking on exposed outcrops. This behaviour makes them relatively easy prey and, as such, at least one adult group member will keep watch for potential predators, emitting a sharp cry if danger approaches.

Dassies are found in a strip bordering the southern edge of the Sahara Desert, extending down the eastern coast into South Africa, and up the west coast into Angola.

HABITAT Mountainous areas and rocky outcrops, occasionally coastal scrub.

DIET A variety of plant matter, including flowers, fruit and bark.

TRACK N° 18

Typically diurnal, with over 20 documented vocalisations – from harsh, almost donkey-like braying to short, high-pitched yaps.

19 | Large-spotted genet
Rooikol muskejaatkat
Genetta tigrina

Large-spotted genets are solitary, nocturnal animals. During the day, they rest in dense clumps of foliage, making them difficult to see.
As agile tree climbers, large-spotted genets spend a fair amount of time in the canopy, and are capable of jumping between tree tops.

Their range extends throughout most of sub-Saharan Africa, with a gap in southwest Africa that includes much of Namibia and South Africa, and part of Botswana.

HABITAT Woodland and forested areas.

DIET Insects and other invertebrates; also reptiles, birds, rodents and occasionally fruit.

TRACK N° 19

A shrill, shrieking wail (similar to that of an injured dog) that may be heard at night.

20 | Mohol galago (Lesser bushbaby)
Nagapie
Galago moholi

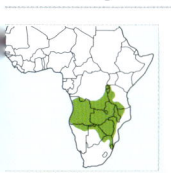

Known for their ability to jump great distances, mohol galagos move about the tree canopy with ease.
　Although these nocturnal animals live in small family groups, they tend to forage alone. Most foraging occurs in trees, but they will sometimes forage on the ground.
　Mohol galagos are common throughout their range, and can be found from the southern parts of central Africa to the northeastern parts of southern Africa.

HABITAT Woodland (acacia, savanna and riverine).

DIET Mostly insects and tree gum; also other small invertebrates.

 TRACK N° 20

A squeaky, high-pitched, baboon-like barking and very loud, siren-like wails. These nocturnal sounds can carry long distances.

Chris & Mathilde Stuart / FLPA

21 | Samango (Sykes') monkey
Samango-aap
Cercopithecus albogularis

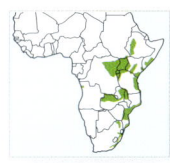

Samango monkeys live in large troops of up to 40 members, which are typically led by one dominant male.
　These monkeys spend most of their time in trees, though they will occasionally forage on the ground.
　Although more reserved than many monkey species, samango monkeys have a range of loud calls, which makes them easy to locate in their dense habitats.
　They have a patchy distribution that ranges from the east coast of South Africa up to Somalia and into parts of central and eastern Africa.

HABITAT Various, from swamps to scrub, though always densely forested.

DIET A variety of plant matter, including fruit and flowers; insects may also be eaten.

 TRACK N° 21

Loud honking and barking calls, often with a guttural tone, may be heard during the day.

Social grooming behaviour is common

22 | Chacma baboon
Kaapse bobbejaan (Bobbejaan)
Papio cynocephalus ursinus

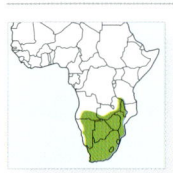

Chacma baboons are the most common and widespread primates in southern Africa.

They are extremely social animals that live in male-dominated troops of up to 100 individuals. They spend much time grooming one another, behaviour that is thought to help forge relationships within the troop, and possibly increase access to food for those doing the grooming.

They forage throughout the day, returning to their sleeping areas before night.

Chacma baboons that have grown accustomed to the presence of humans may climb onto vehicles and attempt to enter houses – in urban areas, they are considered pests.

They are widespread throughout most of southern Africa, extending into the southern parts of central Africa.

HABITAT Varies; from desert, forest and mountain regions, through to suburban areas – almost anywhere that has water and trees.

DIET Omnivorous; a varied diet, ranging from roots and fruit to insects, rodents, birds, snakes and even small antelope. Opportunistic and will attempt to steal any food left unguarded.

TRACK N° 22

Various diurnal sounds: low-volume grunting when foraging; loud grunts, barks and squealing noises when agitated. Adult males make loud 'BAH-who' calls.

23 | Vervet monkey
Blouaap
Cercopithecus pygerythrus

Vervet monkeys are social animals that live in troops of about 20 individuals. Troops have a very clear hierarchical structure, and are led by a dominant male.

Although vervet monkeys may be seen in trees, foraging typically takes place on the ground. Much of their time is spent in clearly demarcated territories.

These monkeys are found throughout most of sub-Saharan Africa, with the exception of central western Africa and the western parts of southern Africa.

HABITAT Acacia and broadleaf woodland; they avoid desert and forested habitats.

DIET Grasses, fruit, acacia leaves, insects, lizards, young birds and even rodents.

TRACK N° 23

Heard during the day, they make a range of agitated, high-pitched, barking sounds with several distinctive alarm calls.

24 | Hippopotamus
Seekoei
Hippopotamus amphibius

These are demure animals that, at least during the day, tend to keep a low profile, often remaining submerged in water. However, during the night, they will venture several kilometres from water to graze.

Hippos typically live in groups of up to 15 individuals, though it is not uncommon to see solitary males.

They are found along major rivers and dams throughout much of sub-Saharan Africa. They are absent from much of southern Africa.

HABITAT Water bodies that are both sufficiently deep to allow submergence and adjacent to adequate grazing pastures.

DIET Mostly grass, but also floating vegetation.

TRACK N° 24

A very deep, booming, repetitive snort; may also make a roaring noise, vaguely similar to that of a lion. May be heard at any time.

19

25 | African elephant
Afrikaanse olifant
Loxodonta africana

A young male

Elephants are highly social animals that live in close-knit herds, which typically consist of a dominant matriarch and her offspring; larger herds will include her female relatives and their offspring. Males are transient among herds.

Their range can cover an area of more than 50 km², though single males may travel much longer distances.

Elephants have an appetite that is proportional to their massive size – they may consume over 300 kg of food and 200 litres of water per day.

They have a patchy distribution throughout sub-Saharan Africa. Their numbers have declined dramatically over the past few decades, largely from loss of habitat – the result of rapidly increasing human populations.

HABITAT Highly variable; from desert through to savanna grassland, and even dense forest. They require leafy trees, from which they browse, and access to water.

DIET Basically leaves and grass. They use their trunk to strip bark from trees and dig for roots. Elephants also eat flowers and fruit, sometimes travelling considerable distances to feast on a newly ripe crop – fruit from the marula tree being a favourite.

TRACK N° 25

During the day, a range of loud trumpeting sounds and very low-pitched rumbles may be heard. Many of their sounds are in ultra-low frequencies.

A black rhino's hook-lipped mouth is suited to browsing

26 | Black rhinoceros
Swartrenoster
Diceros bicornis

Two rhino species occur in Africa: the black rhino and the white rhino. Once plentiful in the region, populations of both species have been decimated through hunting and, more recently, poaching.

The black rhino is smaller than the white rhino, and more solitary. It is distinguished by its pointed upper lip, which is well suited to browsing low shrubs. The white rhino has a more square lip, better suited to grazing.

The horns of both species are highly prized for use in traditional medicine, particularly in Vietnam, although there is no medicinal value in what is, essentially, compacted hair.

Presently, only tiny, isolated rhino populations remain in key conservation areas.

HABITAT Woodland, typically near water sources; however, they avoid thick forest and desert. Their present distribution is mainly determined by protection policies.

DIET The black rhino browses leaves and twigs. By contrast, white rhinos, with their square-shaped mouth, graze on grasses.

TRACK N° 26

A mix of squeals, snorts and growls may be heard during the day. Both species of rhino produce similar sounds.

An adult protects a young herd member

27 | African buffalo
Buffel
Syncerus caffer

Known for their aggresive behaviour, African buffaloes are among the most dangerous animals in the African bush.

They live in very large, male-dominated herds, which may consist of several thousand individuals. Smaller bachelor herds may also be seen. Members will close ranks when they are under attack, and have been seen to drive off an entire pride of lions by rushing at them en masse in order to save one of their own.

Buffaloes are active at the extremes of the day, preferring to rest in the shade during the hottest hours.

They are found throughout the northern parts of sub-Saharan Africa and eastern central Africa. They are largely absent from much of southern Africa.

HABITAT Open savanna and woodland; they avoid areas of low rainfall, which tend to have shorter grasses that are not suitable as food. They need access to water in order to drink their fill at least once a day.

DIET Various long grasses, including those that grow densely in marsh conditions; occasionally reeds and other aquatic plants during the dry season.

TRACK N° 27

A deep, guttural, almost lion-like growl that may be heard during the day.

28 | Blue wildebeest
Blouwildebees
Connochaetes taurinus

Blue wildebeest are most commonly found in male-dominated herds of about 30 individuals. In the Serengeti, however, they can be found in massive groups with over a million members – these large populations are renowned for their mass migrations.

As a result of hunting pressure, many herds are confined to conservation areas.

Blue wildebeest are found throughout the northern parts of southern Africa and in parts of the Great Rift Valley.

HABITAT Savanna; access to drinking water is important.

DIET Various species of grass.

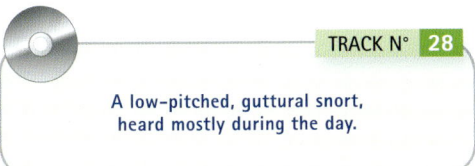

TRACK N° 28

A low-pitched, guttural snort, heard mostly during the day.

29 | Black wildebeest
Swartwildebees
Connochaetes gnou

Black wildebeest live in small, male-dominated herds, though bachelor herds may also be seen.

As a result of low population numbers, they are highly managed and are most often found within conservation areas. The majority of present-day populations are the result of captive-breeding programmes.

Black wildebeest are restricted to central, eastern and northern South Africa.

HABITAT Grassland, savanna; access to drinking water is important.

DIET Several species of grass, occasionally leaves. Need access to drinking water.

TRACK N° 29

A guttural, snorting honk (very similar to their scientific name, 'gnou') that may be heard during the day.

30 | Gemsbok (Oryx)
Gemsbok
Oryx gazella

Gemsbok are generally found in herds of about 30 members, though males may be solitary for part of the year.
In order to conserve energy and keep cool, gemsbok are often inactive during the hottest parts of the day.

They are adapted to arid areas, and are able to survive for long periods without water.

Gemsbok occur in the arid parts of north-western southern Africa.

HABITAT Open grassy plains and arid areas; may venture into woodland.

DIET Grass and leaves; also wild melons, which help them meet their fluid requirements.

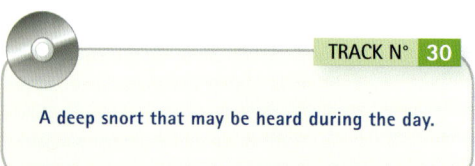

TRACK N° 30

A deep snort that may be heard during the day.

31 | Greater kudu
Koedoe
Tragelaphus strepsiceros

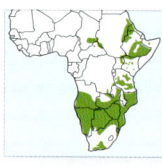

Female greater kudu tend to live in small herds of about 10 individuals, whereas males are usually found alone or in temporary bachelor herds. During the mating season, a single male may be found with a female herd.

Kudu are renowned for their jumping ability – from standing, they can jump several metres into the air.

They are most common throughout the northern parts of southern Africa, with a small range in southernmost South Africa. They have a patchy distribution in eastern Africa.

HABITAT Savanna and woodland habitats.

DIET Leaves and shoots; during the winter months they may eat wild melons and fruit (for their water content).

TRACK N° 31

A very deep snort that may be heard during the day.

Shadow stripes are a distinguishing feature

32 | Burchell's (Plains) zebra
Bontkwagga (Bontsebra)
Equus quagga

These are social animals that live in small family groups, consisting of females, their young and a dominant male. Males will aggressively defend their herd from rivals. These zebras also associate with blue wildebeest and other antelope, probably for purposes of greater communal safety. Zebras defend themselves from predators by kicking and biting, and even lions struggle to take down a full-grown adult.

Zebras rely heavily on good water sources and will typically be found near permanent sources of drinking water.

Their loud, barking call is a sound that is characteristic of the savanna.

Burchell's zebras are found throughout the northern and eastern parts of southern Africa, extending into eastern Africa.

HABITAT Open plains, grassland, savanna, woodland, semidesert and, occasionally, mountainous regions.

DIET Various grasses, particularly the new grass that grows after a fire, or that sprouts quickly after a rain shower. Also occasionally leaves, twigs and roots.

TRACK N° 32

A diurnal call that consists of a distinctive two-noted, almost donkey-like, braying that is repeated several times.

33 | Bushbuck
Bosbok
Tragelaphus scriptus

Bushbuck are usually solitary animals, though they may form loose associations with other bushbuck nearby.

Although their territories are relatively large, their movements are restricted by the accessibility of drinking water.

Bushbuck exhibit predominantly crepuscular and nocturnal habits, though they may be active at any time.

They are found throughout much of sub-Saharan Africa, but are absent from central west Africa, the extreme eastern parts of east Africa and large, western portions of southern Africa.

HABITAT Dense habitats, such as woodland, riverine bush and forests.

DIET Mostly leaves; also shoots, flowers, fruit and grasses.

TRACK N° 33

A low-pitched, dog-like, barking noise that may be heard during the day.

34 | Impala
Rooibok
Aepyceros melampus

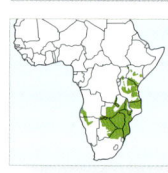

Impala are typically found in herds, separated by gender. Female herds, which also include offspring, usually consist of up to 20 members.

Although males form bachelor herds throughout much of the year, they become territorial and very vocal during mating season.

Impala have a patchy distribution throughout the eastern parts of Africa. They can also be found within a small area on the western Namibia-Angola border.

HABITAT Grassland, savanna and woodland.

DIET Varies seasonally. Predominantly grass in the wet season, and foliage, shoots and seeds in the dry season.

TRACK N° 34

A high-pitched, short, sharp snort, reminiscent of the sound of tearing cloth; heard during the day.

35 | Springbok
Springbok
Antidorcas marsupialis

Springbok typically live in small herds, though if resources are plentiful, herds may increase in size.
  Males have a mating ritual that involves 'pronking', which refers to the action of jumping up and down on stiff legs – an activity that can be quite comical to watch.
  The springbok is a national symbol of South Africa, known for its running speed (up to 100 km/hour) and jumping ability (several metres).
  Springbok are typically diurnal, but may also be active during the crepuscular hours.
  They are restricted to the arid parts of northwestern southern Africa.

HABITAT Dry habitats, from desert scrub to grassy plains.

DIET Preferentially grasses; may browse shrubs and succulents.

TRACK N° **35**

A low-pitched, nasal, gurgling, purring, frog-like sound with the occasional squeaking noise; heard during the day.

36 | Cape fur seal
Kaapse pelsrob
Arctocephalus pusillus

During the breeding season, male fur seals will establish and aggressively defend a territory with a breeding colony. The pregnant females arrive soon afterwards and give birth, and mating takes place six days after the birth.
  Although some land-based predators may attack breeding colonies, the seals' greatest predatory threat comes from great white sharks and killer whales. Fishermen may also persecute fur seals on account of competition for dwindling fish stocks.
  Cape fur seals are restricted to the west and south coasts of southern Africa.

HABITAT When on land, they prefer rocky islands and pebbly, stony beaches.

DIET Mainly fish; occasionally squid, crabs and other crustaceans.

TRACK N° **36**

Loud barking and braying sounds that may be heard during the day.

37 | Egyptian fruit bat
Egiptiese vrugtevlermuis
Rousettus aegyptiacus

Fruit Bats of the genus *rousettus* are the only fruit bats to use echo location. They can thus utilise deep, dark caves, where they form roosts consisting of several thousand individuals.

Egyptian fruit bats are able to carry fruit with one claw and feed while flying. In a single night, they may fly up to 20 km in search of food.

They have a patchy distribution throughout sub-Saharan Africa, and tend to be found in proximity to coastal regions.

HABITAT Caves and heavily wooded areas; they require access to fruit.

DIET Various fruit.

TRACK N° 37

A high-pitched clicking noise that may be heard at night.

38 | African Fish Eagle
Visarend
Haliaeetus vocifer

African Fish Eagles will hunt from perches that overlook water. They are known for swooping dramatically from on high to catch fish.

Although adept hunters, they are capable of stealing fish from other birds.

As with many raptors, young Fish Eagles will display the 'Cain and Abel' phenomenon, whereby a stronger chick will ultimately kill its weaker sibling.

These birds are widespread throughout much of sub-Saharan Africa, but are absent from the drier Horn of Africa region and the desert regions of southern Africa.

HABITAT Large rivers, wetlands and lagoons.

DIET Mostly fish; may also scavenge.

TRACK N° 38

A distinctive, far-carrying, diurnal call, often referred to as the 'sound of Africa'. During breeding season, pairs often duet.

39 | Hadeda Ibis
Hadeda
Bostrychia hagedash

Hadeda Ibises call frequently and loudly, making them a prominent species within their range.

As with most birds that dig for their food, Hadeda Ibises have an acutely developed sense of smell.

They have an extensive distribution across sub-Saharan Africa, but are absent from Namibia, Angola, parts of Botswana and a small area below the Horn of Africa.

HABITAT Woodland, savanna, parkland and around wetlands.

DIET Mainly underground invertebrates, such as worms, though they will occasionally eat insects and small lizards.

TRACK N° 39

A raucous 'ha ha hadeda', from which this species derives its name; may be heard during the day.

40 | Hamerkop
Hamerkop
Scopus umbretta

Hamerkops are well known for their massive nests, made from sticks and twigs, which are built in the forks of very large trees.

They are often seen knocking their prey against rocks, which serves to stun and immobilise it before it is swallowed.

Hamerkops are widespread across the sub-Saharan region, with the exception of a few small areas along the southern west coast region and in the drier areas of Namibia and Botswana.

HABITAT In the vicinity of fresh water, including streams, dams and wetlands.

DIET Mainly frogs and small fish.

TRACK N° 40

Heard during the day; a loud, harsh and excited cackling or laughing call that rises and falls in pitch.

41 | Southern Red Bishop
Rooivink
Euplectes orix

Like most brightly coloured members of this family, the male Southern Red Bishop turns a drab brown colour during winter.

These birds are very territorial in the breeding season, though will gather in huge flocks, with other seed-eaters, over the winter months.

Their nests are elaborately woven from grasses and are normally mounted between two reeds or other supporting plants.

Southern Red Bishops have a patchy distribution in central Africa, with a more extensive range across southern Africa.

HABITAT Traditionally associated with wetlands; they also frequent open farmland and grassland.

DIET Predominantly seeds; will occasionally consume termites.

TRACK N° 41

A series of diurnal churring and swizzling sounds, interspersed with a distinctive sharp, descending whistle.

42 | Egyptian Goose
Kolgans
Alopochen aegyptiaca

Egyptian Geese are largely terrestrial and are commonly seen perching on buildings and in trees.

They are very territorial birds, and males fight aggressively over breeding ground.

Egyptian Geese have been known to steal nests from other birds, including from small raptors.

This species is widespread throughout sub-Saharan Africa, except in coastal regions along the 'bulge of Africa' and in some areas along the west coast of southern Africa.

HABITAT Near bodies of fresh water, such as streams, dams and wetlands.

DIET A variety of plant matter, including seeds and aquatic vegetation.

TRACK N° 42

Harsh or hissing, goose-like honks and grunts that may be heard during the day.

43 | White-faced Whistling Duck
Nonnetjie-eend
Dendrocygna viduata

White-faced Whistling Ducks will occasionally nest in trees, which is unusual for a duck species.

Although they may nest up to 2 km from water for much of the year, during the winter months, when they moult and are rendered flightless, they tend to congregate around bodies of water.

These birds are most active during the morning and evening hours, when they can often be seen wading in shallow water.

White-faced Whistling Ducks can be found across large parts of sub-Saharan Africa; they are absent from parts of central Africa, as well as the western central parts of the continent and the drier parts of southern Africa.

HABITAT Inland fresh water.

DIET Aquatic vegetation.

TRACK N° 43

An airy, high-pitched, three-noted whistle, with each note lower than the previous; may be heard during the day.

44 | Helmeted Guineafowl
Gewone tarentaal
Numida meleagris

Unlike most birds, Helmeted Guineafowl tend not to fly, more often opting to run as a flock. As such, they make easy prey and are often hunted for sport. They have proved a valuable food resource since ancient Roman times and are presently farmed in fairly large numbers for the game-bird industry.

Helmeted Guineafowl often create a lot of noise at dusk, as they settle for the evening.

These birds occur across much of the sub-Saharan region, except in an area of central Africa, extending from the Congo to the western coastal areas.

HABITAT From arid scrub to grassland to forest edges. They roost communally in trees.

DIET Primarily seeds, berries and bulbs, though will occasionally feed on insects.

TRACK N° 44

A typical chicken-like clucking and a short, two-noted, whistled phrase given by territorial males; heard during the day.

45 | Black Crake
Swartriethaan
Amaurornis flavirostra

Black Crakes are the least elusive of the crakes, which are notoriously secretive birds, and are often seen in the open. They are more readily seen throughout the winter months.

As co-operative breeders, all members of the group assist with nesting and raising the young.

Black Crakes are relatively noisy birds that spend much of their time calling.

They occur across sub-Saharan Africa, with the exception of the drier parts of southern Africa and the Horn of Africa.

HABITAT Reed beds and tangled vegetation on the edges of fresh water.

DIET Aquatic vegetation, insects, small fish and frogs.

TRACK N° **45**

An amusing diurnal duet between a breeding pair. One gives a high-pitched chuckle, the other produces a lower-pitched, bubbling murmur.

46 | Blue Crane
Bloukraanvoël
Anthropoides paradiseus

Graceful and attractive, the Blue Crane is the national bird of South Africa.

Although once threatened, population numbers are presently healthy in many farming regions.

Blue Cranes are generally social birds. Mated pairs tend to disperse during breeding season, but return to the same nesting site every year. Males and females will take turns incubating the eggs.

They are almost exclusive to South Africa, but are absent from the Northern Cape and extreme northern parts of the country. There is a small, isolated population in Namibia's Etosha Pan.

HABITAT Grassland.

DIET Bulbs, roots, grains, various insects, fish, frogs and other small creatures.

TRACK N° **46**

A guttural, rolling, almost croaking call – often performed in duet; heard during the day.

47 | African Jacana
Grootlangtoon
Actophilornis africanus

African Jacanas have disproportionately long toes that, combined with their light body mass, allow them to walk on floating plants.

These birds are polyandrous – the female will mate with several males, which then raise the chicks without the female's help. Males are known for carrying their young under their wings and, occasionally, the feet of young birds may be seen sticking out from under their wings.

They are found in most parts of the sub-Saharan region, but are absent from the drier parts of southern Africa.

HABITAT Restricted to fresh water: mostly shallow areas with dense vegetation such as water lilies.

DIET Primarily insectivorous, but may consume plant matter and small fish.

TRACK N° 47

A short, repetitive, high-pitched churring sound, which is repeated in short bursts of up to ten notes; heard during the day.

48 | Spotted Thick-knee
Gewone Dikkop
Burhinus capensis

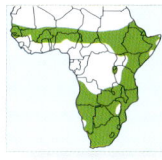

Spotted Thick-knees exhibit predominantly nocturnal and crepuscular behaviour, to which their large, striking yellow eyes are well adapted.

During the day, these birds may be difficult to see because they typically stand motionless in the shade of a tree: a habitat where they are well camouflaged.

Their call may be heard from great distances as breeding pairs and territorial birds interact.

Spotted Thick-knees occur throughout southern Africa, extending up the eastern half of the continent, across the southern end of the Sahara Desert, to the west coast.

HABITAT Wide-ranging: almost any area containing short grass and stony ground.

DIET Mainly insects.

TRACK N° 48

A ghostly, shrill, piercing, whistled call, which varies greatly in pitch and speed; heard during the day.

49 | Crowned Lapwing
Kroonkiewiet
Vanellus coronatus

Most Crowned Lapwings are sedentary, although, during particularly dry periods, they may become nomadic; some populations are migratory.

These birds tend to nest in loose colonies, and typically choose new breeding sites each year. They are well known for their dive-bombing tactics during the nesting season.

Adults have a sharp nail-like spur on the fold of their wings, although it is often difficult to spot.

Crowned Lapwings occur throughout most of southern Africa, extending up the east coast to the Horn of Africa.

HABITAT Restricted to areas of short grassland.

DIET Insects, with a particular affinity for termites.

TRACK N° 49

Their high-pitched, metallic notes and guttural purrs may be heard during the day.

50 | Cape Turtle Dove
Gewone Tortelduif
Streptopelia capicola

Cape Turtle Doves have the most repetitive and monotonous call of all the doves – it is probably the most recognised of the bird calls within its range.

As they are dependent upon surface water, it is not uncommon to see large flocks of Cape Turtle Doves gathered at water sources, particularly in drier regions.

These birds are found from southern Africa, up to the southern half of central Africa, extending up the eastern side of the continent to the Horn of Africa.

HABITAT Treed areas; may also be found in small clumps of trees in open, arid areas.

DIET Predominantly seeds; occasionally fruit and insects.

TRACK N° 50

A distinctive and repetitive 'tur turr-a' sound, from which their name is derived; heard during the day.

51 | Grey Go-away Bird
Kwêvoël
Corythaixoides concolor

Their common name is derived from their alarm call, which sounds like someone shouting 'go away'. These are noisy birds, particularly during the breeding season.

Grey Go-away Birds are very agile in trees and prefer to run, rather than hop, along branches. They owe this agility to their zygodactylous feet (two toes point forward and two point backward).

As cities have begun incorporating more trees, their range has expanded. They are found across the northern parts of southern Africa, extending into the southern parts of central Africa.

HABITAT Dry woodland and savanna, particularly acacia.

DIET Mostly fruit and berries; also flowers, buds and nectar.

TRACK N° 51

Their harsh 'kwe' or 'go-away' call, as well as harsh, guttural contact calls may be heard during the day.

52 | Purple-crested Turaco
Bloukuifloerie
Tauraco porphyreolophus

Albert Froneman

Like most brightly coloured turacos, Purple-crested Turacos have a red pigment (turacin) in their wing feathers. The moment they spread their wings, an explosion of bright, fiery red can be seen, although it's hard to detect when they are perched.

Purple-crested Turacos are typically secretive birds, though they occasionally gather at fruiting trees and become quite raucous.

They have been known to build their nests in termitaria, provided there is adequate camouflage.

These birds are restricted to the far eastern parts of central and southern Africa.

HABITAT Evergreen or riverine forests.

DIET Almost exclusively fruit.

TRACK N° 52

A loud, piercing and resonant 'kok kok kok' that may be heard during daylight hours.

53 | Burchell's Coucal
Gewone Vleiloerie
Centropus burchellii

Burchell's Coucals are known for their gluttonous appetites. They will attack almost any prey that their small size allows. They may even raid the nests of other birds, taking eggs and hatchlings.

These coucals are not particularly graceful and tend to crash-land in trees – a comical sight.

Burchell's Coucals can be found in the far southeastern parts of Africa, extending south and west along the coast of southern Africa.

HABITAT Dense vegetation, such as forest edges, lush gardens and reed beds.

DIET Various, from small mammals and reptiles to nestlings, eggs and insects.

TRACK N° 53

A deep, liquid-bubbling call that slowly descends in pitch; heard during the day.

54 | Spotted Eagle Owl
Gevlekte Ooruil
Bubo africanus

As with most owls, Spotted Eagle Owls are typically nocturnal, spending their days roosting in trees or amidst rocks. At sunset, they fly to a perch where they can begin hunting, which they do very effectively – largely due to their honed senses and near-silent flight.

Spotted Eagle Owls are the largest and most common owls in much of Africa. They are widespread across southern Africa, though may avoid some very dry western areas.

HABITAT Grassland near rocky areas, urban centres and any area with large trees; they avoid dense forest.

DIET Mainly small mammals; also snakes, frogs and other birds.

TRACK N° 54

A soft duet: the male calls a soft, two-noted hoot and the female responds with a slightly higher-pitched, three-noted hoot.

55 | Pearl-spotted Owlet
Witkoluil
Glaucidium perlatum

Pearl-spotted Owlets are known for their false eye-spots, located on the back of the head, which give potential predators the impression that they are always being watched. They are among the smallest owls in Africa.

Although these owls are often active during the day, they hunt mostly at night.

They occur in a narrow belt south of the Sahara Desert, extending down the eastern half of Africa and across the northern parts of southern Africa.

HABITAT Woodland and bushveld; they avoid very dense habitats.

DIET Small rodents, reptiles, birds and insects.

TRACK N° 55
A series of distinctive, piercing, whistles that slowly rise in pitch, culminating in a crescendo of short downward-slurred whistles.

56 | Fiery-necked Nightjar
Afrikaanse Naguil
Caprimulgus pectoralis

Malcolm Schuyl / FLPA

Fiery-necked Nightjars are well suited to their woodland habitat: they sing and hunt from their tree perches and use leaf litter on the forest floor as a camouflage.

Their call is widely known throughout Africa and is heard most frequently during the night, throughout the dry season; they seldom call during the wetter months.

Fiery-necked Nightjars can be found throughout southern Africa, except in the drier western areas.

HABITAT Woodland.

DIET Almost exclusively insects.

TRACK N° 56
A jumbled series of whistles, which sound similar to the phrase 'good Lord deliver us', heard during the night.

57 | Woodland Kingfisher
Bosveldvisvanger
Halcyon senegalensis

These striking looking, blue-backed kingfishers tend to be seen more frequently than many of the other kingfishers, although even they are often heard before they are seen.

Most populations of Woodland Kingfishers tend to be migratory and are only present in southern Africa during the summer months.

Their range extends across much of sub-Saharan Africa, with the exception of the drier areas in the Horn of Africa and the south-western parts of southern Africa.

HABITAT Woodland.

DIET Mainly insects; rarely crustaceans, fish, reptiles, amphibians and small mammals.

TRACK N° 57

A very characteristic piercing 'chip-churrrrrrr', which is repeated monotonously for long periods; heard during the day.

58 | Southern Ground Hornbill
Bromvoël
Bucorvus leadbeateri

Southern Ground Hornbills are social animals that tend to forage in groups. They spend most of their time on the ground, though will roost and take refuge in trees.

These birds are unique in that they walk solely on the tips of their toes.

Southern Ground Hornbills are found throughout much of southeastern central Africa, extending westwards in the northern sections of southern Africa, as well as southwards along the eastern coast.

HABITAT Grassland, savanna, woodland and occasionally forest.

DIET Insects, reptiles, small mammals, occasionally small birds or nestlings.

TRACK N° 58

Their deep, booming, staccato call with a series of deep, pounding, booming notes may be heard during the day.

59 | Southern Yellow-billed Hornbill
Geelbekneushoringvoël
Tockus leucomelas

Southern Yellow-billed Hornbills spend much of their time foraging on the ground. When threatened, however, they will take refuge in trees.

These social hornbills often call together as a group – if an individual begins calling, others usually join in.

They can be found in a narrow strip along the west coast of central Africa, extending eastwards to Mozambique; they are also found in the northern and eastern regions of southern Africa.

HABITAT Broadleaf and acacia woodland.

DIET Mostly insects.

TRACK N° 59
During the day, a series of staccato 'tuck' notes may be heard. These notes are repeated fairly rapidly, but never change into a double-noted song.

60 | Green Wood-hoopoe
Rooibekkakelaar
Phoeniculus purpureus

Green Wood-hoopoes give entertaining displays, in which members in a group roll back and forth on a branch while giving their raucous, cackling call.

These birds are co-operative breeders that live in family groups with only one breeding pair. Throughout the incubation period, females will remain in the nest with the young; other flock members will bring food to the nest.

Green Wood-hoopoes are located throughout much of sub-Saharan Africa, with the exception of western central Africa, the drier regions in the Horn of Africa and desert regions in southern Africa.

HABITAT Mainly woodland, savanna with scattered trees, wooded gardens and parklands.

DIET Mostly insects.

TRACK N° 60
A loud, raucous, cackling sound with an almost bubbling quality; heard during the day.

61 | African Hoopoe
Hoephoep
Upupa africana

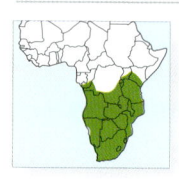

African Hoopoes are generally not shy and, as a result, are commonly seen. These birds are, at least partially, migratory, so that they are more prevalent locally in the spring and summer months.

African Hoopoes have disproportionately short legs relative to other ground-based foraging birds, giving them a characteristic slightly comical stance.

They are widespread throughout southern Africa, extending north into the southern parts of central Africa.

HABITAT Open habitats, including savanna, gardens and parks, and light woodland.

DIET Mostly insects.

TRACK N° 61

During the day, a two-noted 'hoop-hoop' or a three-noted 'hoop-hoop-hoop' call may be heard.

62 | Black-collared Barbet
Rooikophoutkapper
Lybius torquatus

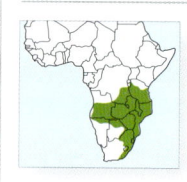

Using their powerful bills, Black-collared Barbets hollow out nest cavities within dead wood.

These birds have developed elaborate greeting rituals, which involve loud duets combined with bobbing, swaying and wing flicking.

A rare yellow variation exhibits a yellow face and throat.

Black-collared Barbets inhabit the eastern regions of central and southern Africa, extending westwards through the northern parts of southern Africa and the southern parts of central Africa.

HABITAT Woodland and riverine bush, avoiding dry acacia habitats and montane regions.

DIET Predominantly fruit; occasionally insects and larvae.

TRACK N° 62

Various calls are heard during the day, the most common of which is a duet that is best described as 'two-puddly'.

63 | Crested Barbet
Kuifkophoutkapper
Trachyphonus vaillantii

Crested Barbets excavate cavities in dead trees in which to make their nests.
They are relatively territorial and tend to drive away most other species – particularly those that may compete for nesting spots.

Unlike other barbets, which only occasionally venture to the ground, Crested Barbets are more terrestrial and will frequently feed on the ground.

These are the only barbets in the region to have a crest.

They live throughout northeastern southern Africa, extending as far west as Angola.

HABITAT Dry woodland habitats, occasionally gardens and parks.

DIET Mostly insects; also fruit.

TRACK N° 63
A long, drawn-out, purring or churring call, persisting at a constant pitch for long periods throughout the day. Females sometimes respond with an agitated chattering.

64 | Rufous-naped Lark
Rooineklewerik
Mirafra africana

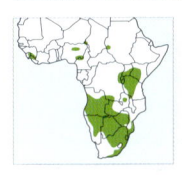

Song is an important aspect of the Rufous-naped Lark's behaviour. Displaying birds may call for hours from a raised perch.

Although their calls can be highly variable, they usually maintain similar structures. Rufous-naped Larks may also mimic a variety of other bird calls.

These are by far the most obvious larks in their range, owing to their wide distribution, prominent perching and vocal behaviour.

Rufous-naped Larks have a patchy distribution in sub-Saharan Africa, with the largest range extending from southwestern central Africa to southeastern southern Africa.

HABITAT Open grassland; also woodland, scrubland and farmland.

DIET Insects and seeds.

TRACK N° 64
A two- to three-noted 'trillee-trilloo' or 'tree-lee-loo', which is often repeated incessantly during the day.

65 | Red-capped Robin-chat
Nataljanfrederik
Cossypha natalensis

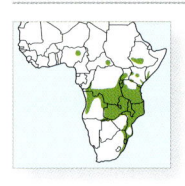

Red-capped Robin-chats are known for their vocal abilities. They are able to mimic a wide range of sounds, including non-bird calls (e.g. dogs barking). It is not uncommon for an individual to mimic the calls of more than a dozen different bird species.

These birds enjoy bathing and, when free-standing water is not available, they will even bathe in the dew from wet leaves.

Red-capped Robin-chats are found across the central and eastern parts of Africa, extending into southern Africa via a narrow belt along the eastern side of the continent.

HABITAT Dense forest, woodland, riverine bush and lush gardens.

DIET Insects and berries.

TRACK N° 65
The song consists of a wide variety of tuneful notes and many mimics, but a two-noted guttural sound is unique to this species.

66 | Bokmakierie
Bokmakierie
Telophorus zeylonus

Bokmakieries are sociable songbirds with a colourful vocal repertoire. Male and female pairs will duet, often from a prominent perch; through the use of calls, these pairs proclaim and maintain their territory throughout the year.

Bokmakieries build their grass nests in trees and shrubs.

They are restricted to South Africa, the eastern sections of Namibia and a narrow strip of Angolan coastline.

HABITAT Open habitats, particularly scrub and grassland with scattered bushes; also gardens.

DIET Mainly insects, but will consume small birds, amphibians and snakes; occasionally berries.

TRACK N° 66
Throughout the day, a wide range of piercing whistles, hoops, pops and trills (usually sung as a duet between males and females) may be heard.

67 | Grey-headed Bushshrike
Spookvoël
Malaconotus blanchoti

Grey-headed Bushshrikes are the largest members of the African shrikes.

With their large, powerful, hooked bill, they are capable of killing and dismembering a variety of prey, including mice, frogs and small snakes. They are known to impale their prey on sharp objects such as thorns or fence poles, thus creating a 'larder'.

They occur in a belt below the Sahara Desert, extending down the eastern coast of Africa and across the northern and southeastern parts of southern Africa.

HABITAT Woodland, riverine forest and the edges of dense forest; occasionally gardens.

DIET Mostly large insects; occasionally small vertebrates.

TRACK N° 67

A long, ghostly, drawn-out hoot, sometimes as a duet by a male-female pair, in addition to other short, harsh, grating sounds; heard during the day.

68 | Southern Masked Weaver
Swartkeelgeelvink
Ploceus velatus

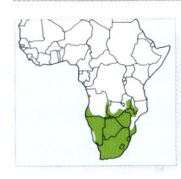

Male Southern Masked Weavers weave their nests onto the end of a stripped tree branch. The first few pieces of grass are elaborately tied to the branch in a secure, formal knot to prevent the nest from falling out of the tree.

Males are very brightly coloured in summer, but turn a drab greenish-brown in winter.

Southern Masked Weavers can be found throughout most of southern Africa, extending into central Africa.

HABITAT A wide range that includes trees; not found in dense bush or forest.

DIET Insects and a variety of plant matter, including seeds and fruit.

TRACK N° 68

Various swizzles, with a characteristic rising, churring sound, followed by a descending nasal sigh; heard during the day.

69 | Common barking gecko
Kleinblafgeitjie
Ptenopus garrulus

Common barking geckos live alone in burrows that are often extensive, and which they dig out of firm sand.

These reptiles use sound for a range of communication, including attracting mates and defending territories. Males will typically call from within their burrows, and are most often heard on summer evenings; they may also call on very overcast days.

Common barking geckos have a sticky tongue, which they use to catch prey.

They are limited to the western and central regions of southern Africa.

HABITAT   Ranges from arid mountain peaks to desert regions.

DIET   Almost exclusively small insects.

TRACK N° 69

A nocturnal, high-pitched, almost cricket-like bark, from which this species derives its name.

70 | Mole cricket
Molkriek
Gryllotalpa africana

Male mole crickets produce a loud buzzing sound by rubbing their forewings over one another. They typically call from a particular area at the entrance to their burrow, which they shape in a way that amplifies their call, while making the source difficult to locate.

These insects have large, shovel-like forelimbs, which are highly adapted for digging – their burrows may be up to 1 m deep. Mole crickets often cause damage to grass root systems where they burrow.

They are widespread throughout sub-Saharan Africa, although they tend to avoid very dry areas.

HABITAT   Grasses and moist soils.

DIET   Grasses.

TRACK N° 70

A churring noise that is heard most often at night.

71 | Cicada
Sonbesie
Cicadidae Family

Male cicadas produce sound from two tight membranes (tymbals), which cover cavities on the base of their abdomen. The muscles that activate these membranes can contract at over 400 times per second, allowing for the production of relatively high frequencies.

One calling male will typically incite others to join and, in large numbers, they can produce a deafening sound.

There are approximately 140–150 cicada species in southern Africa, with species still being discovered. They are widespread throughout sub-Saharan Africa.

HABITAT Woodland and forested areas.

DIET Plant sap.

TRACK N° 71
Males make a piercing, vibrating call that may be heard at any time of the day or night.

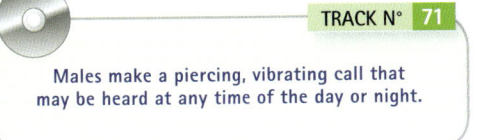

72 | Clicking stream frog
Klik-stroompadda
Strongylopus grayii

Clicking stream frogs, also known as Gray's stream frogs, have a flexible breeding season, which is adjusted according to variations in rainfall.

Although not abundant, their numbers are high where habitat is suitable, as many individuals can co-exist in a small area.

Males will call from well-concealed positions, typically beneath vegetation.

Clicking stream frogs are largely restricted to the southern and eastern parts of South Africa, in addition to Lesotho and Swaziland.

HABITAT Typically scrubby vegetation near water sources.

DIET Predominantly arthropods.

TRACK N° 72
A series of soft metallic clicks, similar to dripping water, which may be heard at any time of the day or night.

73 | Guttural toad
Gewone skurwepadda
Amietophrynus gutturalis

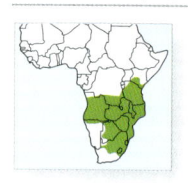

Guttural toads, also known as African common toads, are often found in suburban gardens, where they may keep residents awake at night with their loud croaking. They are particularly noisy during the breeding season.

Guttural toads can be found throughout much of eastern, central and southern Africa, but are largely absent from more arid regions, like Namibia and southwestern South Africa.

HABITAT   Near permanent or semipermanent bodies of water in grassland or savanna; common in gardens.

DIET   Insects.

TRACK N° 73
A typical frog-like croak, which starts out deep and rises in pitch; heard at night.

74 | Raucous toad (Ranger's toad)
Lawaaiskurwepadda
Amietophrynus rangeri

Raucous toads call most frequently during the spring and summer months. Throughout the breeding season, males compete vocally with each other, as well as with male guttural toads.

Raucous toads regularly establish themselves in moist areas near houses where, during the evenings, they can easily catch insects that are drawn to the house lights.

These toads are mostly found in the eastern half of South Africa, extending along the south coast to the Western Cape.

HABITAT   Running water, such as rivers and streams, within grassland; common in gardens.

DIET   Insects.

TRACK N° 74
A loud, raucous, duck-like 'kgwe-kgwe-kgwe', which may continue for hours during the night.

75 | Bubbling kassina
Silwerbruin–vleipadda
Kassina senegalensis

Bubbling kassinas are very small frogs: even adults do not exceed 5 cm in length. Males begin calling towards evening, and move closer to water. By the time it's dark and they are concealed near the water source, they start calling as a collective. When they call in large numbers, their chorus makes a bubbling sound.

Bubbling kassinas live throughout sub-Saharan Africa, except in the Congo basin, the Horn of Africa and the arid southwest.

HABITAT Grassland surrounding vleis, marshes and pans.

DIET Insects.

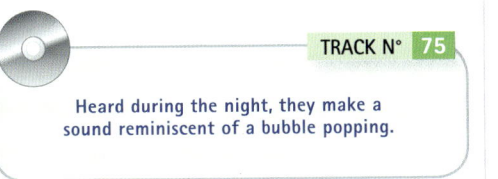

TRACK N° 75

Heard during the night, they make a sound reminiscent of a bubble popping.

76 | Banded rubber frog
Gomlastiekpadda
Phrynomantis bifasciatus

Leonard Hoffmann / IOA

Banded rubber frogs have a distinct coloration, which – in nature – often indicates toxicity, thereby warding off many potential predators.

Although traditionally ground dwellers, these frogs are capable tree climbers.

Males call from a water source where they have concealed themselves amidst vegetation or rocks.

Banded rubber frogs are found throughout the northern parts of southern Africa, extending into northeastern South Africa.

HABITAT Various semi-arid to subtropical environments, including woodland and grassland.

DIET Mostly insects.

TRACK N° 76

A loud, piercing 'trrrrrrrrrrrrrrr' trill sound, lasting several seconds; it is repeated at intervals throughout the night.

INDEX

	PAGE No	TRACK No
African buffalo	22	27
African elephant	20	25
African Fish Eagle	28	38
African Hoopoe	40	61
African Jacana	33	47
African wild dog	9	04
Banded mongoose	13	13
Banded rubber frog	47	76
Bat-eared fox	10	07
Black-backed jackal	9	05
Black-collared Barbet	40	62
Black Crake	32	45
Black rhinoceros	21	26
Black wildebeest	23	29
Blue Crane	32	46
Blue wildebeest	23	28
Bokmakierie	42	66
Brown hyaena	11	09
Bubbling kassina	47	75
Burchell's Coucal	36	53
Burchell's (Plains) zebra	25	32
Bushbuck	26	33
Cape fur seal	27	36
Cape porcupine	12	11
Cape Turtle Dove	34	50
Chacma baboon	18	22
Cheetah	8	03
Cicada	45	71
Clicking stream frog	45	72
Common barking gecko	44	69
Crested Barbet	41	63
Crowned Lapwing	34	49
Dassie (Rock hyrax)	16	18
Egyptian fruit bat	28	37
Egyptian Goose	30	42
Fiery-necked Nightjar	37	56
Gemsbok (Oryx)	24	30
Greater kudu	24	31

	PAGE No	TRACK No
Green Wood-hoopoe	39	60
Grey Go-away Bird	35	51
Grey-headed Bushshrike	43	67
Guttural toad	46	73
Hadeda Ibis	29	39
Hamerkop	29	40
Helmeted Guineafowl	31	44
Hippopotamus	19	24
Honey badger	12	10
Impala	26	34
Large-spotted genet	16	19
Leopard	7	02
Lion	6	01
Meerkat (Suricate)	14	15
Mohol galago (Lesser bushbaby)	17	20
Mole cricket	44	70
Pearl-spotted Owlet	37	55
Purple-crested Turaco	35	52
Raucous toad (Ranger's toad)	46	74
Red bush squirrel	15	16
Red-capped Robin-chat	42	65
Rufous-naped Lark	41	64
Samango (Sykes') monkey	17	21
Side-striped jackal	10	06
Slender mongoose	14	14
Southern Ground Hornbill	38	58
Southern Masked Weaver	43	68
Southern Red Bishop	30	41
Southern Yellow-billed Hornbill	39	59
Spotted Thick-knee	33	48
Spotted hyaena	11	08
Spotted Eagle Owl	36	54
Springbok	27	35
Tree squirrel	15	17
Vervet monkey	19	23
White-faced Whistling Duck	31	43
Woodland Kingfisher	38	57
Yellow mongoose	13	12